NUMBER 127

ELECTRIC WELDING

CONTENTS

British Library Cataloguing-in-Publication Data
A catalogue record for this book is available from the
British Library

Metal Work

Metalworking is the process of working with metals to create individual parts, assemblies, or large-scale structures. The term covers a wide range of work from large ships and bridges to precise engine parts and delicate jewellery. It therefore includes a correspondingly wide range of skills, processes, and tools. The oldest archaeological evidence of copper mining and working was the discovery of a copper pendant in northern Iraq from 8,700 BC, and the oldest gold artefacts in the world come from the Bulgarian Varna Necropolis and date from 4450BC. As time progressed, metal objects became more common, and ever more complex. The need to further acquire and work metals grew in importance. Fates and economies of entire civilizations were greatly affected by the availability of metals and metalsmiths. The metalworker depends on the extraction of precious metals to make jewellery, buildings, electronics and industrial applications, such as shipping containers, rail, and air transport. Without metals, goods and services would cease to move around the globe with the speed and scale we know today.

One of the more common types of metal worker, is an iron worker – who erect (or even dismantle) the structural steel framework of pre-engineered metal buildings. This can even stretch to gigantic stadiums and arenas, hospitals, towers, wind turbines and bridges. Historically ironworkers mainly worked with wrought iron, but today they utilize many different materials including ferrous and non-ferrous metals, plastics, glass, concrete and composites. Ironworkers also unload, place and tie reinforcing steel bars (rebar) as well as install post-tensioning systems, both of which give strength to the concrete used in piers, footings, slabs, buildings and bridges. Such labourers are also likely to finish buildings by erecting curtain wall and window wall systems, pre-cast concrete and stone, stairs and handrails, metal doors, sheeting and elevator fronts – performing any maintenance necessary.

During the early twentieth century, steel buildings really gained in popularity. Their use became more widespread during the Second World War and significantly expanded after the war when steel became more available. This construction method has been widely accepted, in part due to cost efficiency, yet also because of the vast range of application – expanded with improved materials and computer-aided design. The main advantages of steel over wood, are that steel is a 'green' product, structurally sound and manufactured to strict specifications and tolerances, and 100% recyclable. Steel also does not warp, buckle, twist or bend, and is therefore easy to modify and maintain, as well as offering design flexibility. Whilst these advantages are substantial, from aesthetic as well as financial points of view, there are some down-sides to steel construction. It conducts heat 310 times more efficiently than wood, and faulty aspects of the design process can lead to the corrosion of the iron and steel components – a costly problem.

Sheet metal, often used to cover buildings in such processes, is metal formed by an industrial process into thin, flat pieces. It is one of the fundamental forms used in metalworking and it can be cut and bent into a variety of shapes. Countless everyday objects are constructed with sheet metal, including bikes, lampshades, kitchen utensils, car and aeroplane bodies and all manner of industrial / architectural items. The thickness of sheet metal is commonly specified by a traditional, non-linear measure known as its gauge; the larger the gauge number, the thinner the metal. Commonly used steel sheet metal ranges from 30 gauge to about 8 gauge. There are many different metals that can be made into sheet metal, such as aluminium, brass, copper, steel, tin, nickel and titanium, with silver, gold and platinum retaining their importance for decorative uses. Historically, an important use of sheet metal was in plate armour worn by cavalry, and sheet metal continues to have many ornamental uses, including in horse tack. Sheet metal workers are also known as 'tin bashers' (or 'tin knockers'), a name derived from the hammering of panel seams when installing tin roofs.

There are many different forming processes for this type of metal, including 'bending' (a manufacturing process that produces a V-shape, U-shape, or channel shape along a straight axis in ductile materials), 'decambering' (a process of removing camber, or horizontal bend, from strip shaped materials), 'spinning' (where a disc or tube of metal is rotated at high speed and formed into an axially symmetric part) and 'hydroforming.' This latter technique is one of the most commonly used industrial methods; a cost-effective method of shaping metals into lightweight, structurally stiff and strong pieces. One of the largest applications of hydroforming is in the automotive industry, which makes use of the complex shapes possible, to produce stronger, lighter, and more rigid body-work, especially with regards to the high-end sports car industry.

One of the most important, and widely incorporating roles in metalwork, comes with the welding of all this steel, iron and sheet metal together. 'Welders' have a range of options to accomplish such welds, including forge welding (where the metals are heated to an intense yellow or white colour) or more modern methods such as arc welding (which uses a welding power supply to create an electric arc between an electrode and the base material to melt the metals at the welding point). Any foreign material in the weld, such as the oxides or 'scale' that typically form in the fire, can weaken it and potentially cause it to fail. Thus the mating surfaces to be joined must be kept clean. To this end a welder will make sure the fire is a reducing fire: a fire where at the heart there is a great deal of heat and very little oxygen. The expert will also carefully shape the mating faces so that as they are brought together foreign material is squeezed out as the metal is joined. Without the proper precautions, welding and metalwork more generally can be a dangerous and unhealthy practice, and therefore only the most skilled practitioners are usually employed.

As is evident from this incredibly brief introduction, metalwork, and metalworkers more broadly, have been, and still are – integral to society as we know it. Most of our modern buildings are constructed using metal. The boats, aeroplanes, ships, trains and bikes that we travel on are constructed via metalwork, and mining, metal forming and welding have provided jobs for thousands of workers. It is a tough, often dangerous, but incredibly important field. We hope the reader enjoys this book.

CHAPTER I

ELECTRIC WELDING PROCESSES

Although the electric welding process passed out of the experimental into the practical stage some years ago, electric welding is still a rather vague subject to most mechanics. Electric welding, however, plays an important part nowadays in the manufacture of a great many articles, and several companies have been formed which devote their entire attention to the manufacture of articles in which electric welding is an integral part of the manufacturing process. Without the process of electric welding, many of these products would have to be manufactured in an entirely different way, and in many cases at a greatly increased cost.

There are at least five distinct processes of electric welding in use at the present time. These processes are commonly known as the Zerener, the Benardos, the Strohmenger-Slaughter, the La Grange-Hoho, and the Thomson processes.

The Zerener Process

In the first process mentioned above, the Zerener process, perhaps more commonly known as the electric blow-pipe method, an electric arc is drawn between two carbon electrodes. This arc is then caused to impinge upon the metal surfaces to be welded by means of an electro-magnet. This welding system was introduced by Dr. Zerener of Berlin, Germany, some twenty years ago. No current passes through the work in this case.

The Zerener system, as well as all arc-welding systems, is based upon the fact that when two rods of carbon, connected by suitable means to the poles of a dynamo or to the terminals of current supply cables, are brought into contact, a flame is caused to play between them, this flame being known as an arc. Variations in the gap or distance between the carbon electrodes, or the interposition of resistances of varying intensity, increase or decrease, as the case may be, the amount of current passing through the electrodes, and thus alter the size of the flame or arc.

. In an improvement on this method, known as the Voltex process, the carbons contain a small percentage of metallic oxides—oxide of iron, for instance—which is converted by the heat generated into its metallic form and then vaporized. The vapor tends to increase the size of the arc and minimizes or prevents the carbonization of the work by the carbon of the electrodes at the welding point.

The various systems of electric arc welding are especially valuable when the parts to be welded must, after welding, retain their original positions or relationship with reference to each other. A crack in a machine part illustrates such a case. These methods are also applicable when making an abrupt joint between two plates, filling up holes in castings and generally for any work in which it is necessary to add metal to form the joint.

In arc welding the temperature of the arc is practically impossible of control; to avoid melting, then, even in cases where melting is unnecessary, which not often is the case, it would be compulsory to remove the work or arc at the very instant welding heat was attained, a point not readily determined, having regard to the intense light at the arc necessitating the use of almost opaque spectacles, which not only obscure the vision, but result in severe eye-strain. The heat generated is so intense that it is necessary also to guard the hands and face to avoid burning.

In the case of the iron alloys, the arc method is often open to the objection that it demands infallibility on the part of the operator. While a good weld is almost as good as original continuity, the fact that this is only possible when the pieces are heated to a definite temperature, renders the method less satisfactory than the Thomson process yet to be described. From the point of view of perfection of product, the process is lacking, for the reason that the temperature of the arc—between 5000 and 7000 degrees F.—is far in excess of the melting-point of the iron alloys, and is extremely difficult to control.

The Benardos Process

The Benardos process is also based upon the use of the electric arc, the characteristic principle of this process being that the electric arc is drawn directly between the metal to be welded, which itself forms one electrode for the electric current, and the carbon electrode, which forms the other terminal of the circuit. In this process, the pieces of metal to be welded are melted on their faces together with a small iron rod which acts as a kind of solder and flows in between the two surfaces to be joined together by the welding process. This system, if properly adapted to the work to be done, and with a plant well designed for generating, distributing and regulating the current, is practical, simple and effective. The quantity of current used depends on the thickness to be welded and may, in ordinary practice, range from 200 to 500 amperes. The arrangement of using the metal to be welded as one electrode for the electric circuit makes it possible to obtain a great amount of heat in the weld.

In the Benardos process, the direction of flow of the electric current may be in either direction, but, as a rule, the work or metal electrode is the positive and the carbon electrode the negative pole. In case the flow is reversed, the arc will be shorter and the carbon from the electrode is more liable to enter the weld, thereby hardening the material and rendering it brittle. In commencing the weld, the carbon electrode is brought into contact with the work, thus causing an electric current to flow, but is quickly withdrawn, introducing a resistance which produces an arc of high temperature.

In all arc-welding systems, it is difficult to absolutely prevent the introduction of carbon into the work. An improvement, known as the Slavianoff system, in which a very small arc is used, and which, for that reason, is much slower, prevents, to a large extent, the introduction of carbon into the work, and is therefore preferred, particularly for small work. In this process the electrode is of the same metal as the work.

When a number of welding machines working according to the Benardos system are employed, it is necessary that the current be supplied in such a manner that one machine will not affect the arc of another. This is effected very simply by generating in a compound-wound dynamo of ample capacity, and the machine should be slightly over-, rather than under-, compounded. By this arrangement an increase of load does not lower the voltage. In a well-designed machine the voltage scarcely varies, provided the engine driving it is efficient to maintain its speed. The arcs are arranged in parallel, and each arc is provided with a regulator to adjust the current to the work to be done. The rod of carbon forming the negative electrode is fastened in an insulated holder of light construction. The workman holds this in his hand, strikes the arc by placing the carbon in contact with the work, and manipulates it so as to spread the arc and heat the work at and near the point to be welded with what is described as a soaking heat. When the welding heat is attained, the work is hammered or not according to circumstances. Screens with colored glass windows are used to protect the eyes and skin of the workman from the effect of violet rays.

The Strohmenger-Slaughter System

The Strohmenger-Slaughter system, also an arc-welding system, may be worked with either direct or alternating current. It is generally assumed that the alternating current is preferable for this purpose. The voltage need not be very high and the amount of the current within limits is not important. Successful welding has been carried out with 85 volts used with direct current and 220 volts with alternating current. The quantity of the current depends upon the nature of the work. The parts to be welded are placed in the required position and an electrode is laid upon and along the welding line. This electrode consists of a soft iron rod covered all over except at the extreme ends with a flux suitable for the metal to be welded. Then the work and one end of the electrode are brought into contact, causing, by a series of arcs along the welding line, the electrode to melt and to coat the weld with the flux, thereby preventing oxidation. The flux will flake off when the metal cools. It is claimed that this system is used successfully in the welding of rails and in other repairs by building up worn places, but it is not as generally known as the other systems.

The La Grange-Hoho Process

The La Grange-Hoho system, commonly known as the "water pail" forge, is distinctly different from all other processes in principle as well as in its practical application, and is, properly speaking, only a heating process replacing the blacksmith's fire. A wooden tank is

filled with a bath containing a solution of borax and potassium car-
bonate in water. In this bath, the positive electrode of the electric
circuit is placed. The negative electrode is connected to the metal
to be forged or welded, which is then also immersed in the fluid in
the tank. The metal is then permitted to remain in the fluid until
it has reached the welding temperature. The object to be welded
is then removed and the actual forging or welding process is carried
out in the usual manner on the anvil under a hammer. Strictly
speaking, therefore, this is not an electric welding process, but merely
an electric heating process for bringing the metal to a welding heat.

The Thomson Process

In the Thomson process, also known as the incandescent or resistance process, the metals to be welded are brought into intimate contact, being usually held closely together by metal clamps actuated
by springs so as to permit a permanent pressure on the parts even
when the metal at the welding surfaces commences to melt. By this
contact, the parts to be welded complete an electric circuit, and the
resistance at the points of contact between the metals produces a
welding temperature in a very few seconds, at the same time as the
two metals are, by the spring-actuated clamps, forced together automatically, and welded. A distinct feature of this electric welding
process is that the interior is raised to a welding temperature before
the surface reaches that heat. When heated in the forge for welding,
the opposite conditions take place. In the process of electric welding,
if the exterior surfaces weld, the operator is sure that the interior
is also welded, since it must, by necessity, be of a somewhat higher
heat. With ordinary forge welding the surfaces may present a perfect
weld and still cover an imperfect joint inside.

A few years ago it was thought that electric welding would be
practical only for very small objects, on account of the high amper-
age required, but since that time the process has been developed so
that it is now possible to electrically weld parts of considerable size.
The process is particularly suited for the manufacture of automobile
and bicycle parts, carriage hardware, and mechanics' tools of various
descriptions.

When, as mentioned above, the two parts to be welded have been
placed against each other, in the electric circuit, which heats the
metal at the juncture to a molten state, the separate parts will be
united into one piece in such a manner that the joint is practically
imperceptible, but at first a burr or upset is produced around the
welded surface, composed of the expelled oxidized and otherwise
inferior metal. This oxidation is, of course, removed, and then a
perfect joint is the result.

One very important question in regard to electric welding, and for
that matter any other process for joining metallic parts, is whether
the joint is sound. Experiments and tests, as well as use of electrically welded joints, have unquestionably demonstrated its reliability. In the case of electric welding, the great variety of parts

so joined has shown, beyond doubt, that the joint is practically as sound as the solid sections in the parts so joined. Very commonly the parts which are welded by the electric process are subjected to abuse and rough handling, or to heavy stresses. Especially is this so in automobile work. The results obtained have been so satisfactory as to place the art among the most useful of the applications of electricity.

In this connection, it may be well to mention that the Thomson process, while originally an American invention, has also received considerable attention in England. A writer in the *London Times* some time ago, called attention to the fact that the system has caused a complete revolution in existing methods of manufacture in many industries, and that electric welding had created some entirely new manufactures. As to the reliability of these joints, this writer also mentioned that tests had been carried on regarding the comparative strength of electric and ordinary forged welds, and that these tests show that while the ordinary forge weld of iron bars shows an average strength of 89.3 per cent, as compared with the strength of the solid, electrically welded joints show a strength of 91.9 per cent.

In giving a summary of the advantages which can, with propriety, be claimed for the electric welding process, the following may be stated as being the most important: Finished or nearly finished work may be welded and repaired without damage; the welding operation can be closely watched as it proceeds, and faulty welds prevented; the process is carried out with great rapidity, occupying only a few seconds, and in small work it is performed almost instantaneously; and, finally, impurities are expelled from the joint, and a perfectly homogeneous weld is obtained. The cost for the generation of heat, generally speaking, is probably the same for forge and electric welding, but with the electric process the cost of labor is greatly reduced.

In the following chapters the more important of these welding processes will be dealt with in detail, the methods connected with each being explained and the advantages pointed out.

CHAPTER II

ELECTRIC RESISTANCE PROCESS OF WELDING

In a paper read before the American Society of Mechanical Engineers by Mr. W. A. Hodges, the most important features of the electric resistance process—also known as the Thomson process—of welding were pointed out. This process of welding and heating, sometimes called the incandescent process, as distinguished from the electric arc process, consists in causing a heavy current of electricity to pass across the joint at the lowest voltage which will drive the current through the pieces to be welded, to bring the metal at the junction up to a welding heat; at the same time the pieces are pressed together to make a complete union or weld. The pieces to be welded together when clamped in a welding machine complete an electrical circuit, but are inadequate to carry the heavy current passing through them without heating; as the heat increases the resistance also increases, and the union or weld is thereby accelerated, but the volume of current is decreased; or, in other words, as the temperature increases the current volume usually decreases, a greater volume of current being used at the beginning than at the end of the heat effect. The heat is confined to the metal between the jaws, and a welding heat is reached so quickly that there is very little time for waste through radiation or conduction by adjacent cold metal; therefore practically all the heat is consumed in useful work, and the pieces, not having been heated except at the joint, are not distorted, or even discolored.

The electric resistance process of welding and heating is distinguished from all other processes by the fact that the heat is generated in the metal itself, uniformly over the section, while by all other processes the contrary is true; the heat is applied to the exterior, and is conducted into the interior. Absolute control of the heat is obtained; the pieces are heating in full view of the operator, and, if due to uneven contact of the pieces, rust, or scale at one or more points of contact, or other irregularity, the pieces heat unevenly, with a tendency to burn at one or more points, the heating effect can be instantly stopped until radiation restores the heat equilibrium, when the heating can instantly be resumed, and all danger of flaws avoided. Various degrees of temperature can be obtained and retained for any length of time. Usually no flux is required.

Applications of the Process

This process is employed principally in the following classes of welding operations: Butt welding, end to end; whole abutting surfaces, of nearly same cross-section, welded together; tee and angle welding, in the form of a letter T or L; cross welding, in the form of an X; lap welding, overlapping and squeezing or mashing together; seam welding, either by abutting, or overlapping the edges of sheet metal; spot welding, instead of riveting, practically one spot at a time; point welding, surfaces in contact only at a raised point or multiplicity of points, at which the current is confined, welding only at these points, usually in sheet metal; and ridge welding, where instead of a point, a ridge is employed, across which a weld is made.

The process is especially well adapted to duplicate work where as large an output as possible per operator is required, and the wagon, carriage, automobile and bicycle industries, with tools, wire, pipe, tubing and a large line of miscellaneous and special work, provide the field for the process. It is particularly applicable to new work rather than to repairs, although some repair work is done in iron and steel, but the process is not applicable for the repair of broken or defective castings unless of such metals as brass of simple forms. Practically all metals can be welded, also all sorts of steel and many alloys, as well as many combinations like carbon steel to mild steel, nickel and brass to platinum, cast steel to machine steel, malleable iron to steel and a great variety of other combinations. In the last few years there has been a great development in the manufacture of sheet metal articles and electric spot welding has been found a much cheaper and better method than riveting, especially with the lighter gages of sheet.

Rust is an insulator for low voltage and should be removed at those parts of the pieces which come in contact with the electrodes. In butt welding the two pieces, if of the same metal, should have practically the same cross-section at the joint, and when a larger piece is to be welded to a smaller, the end of the larger piece should be reduced to the section of the smaller for a length depending on the section or diameter. The upsetting of the ends together to make a butt weld causes the joint to be enlarged, forming a burr, fin or swelling; if objectionable, this can be removed under a press or hammer while the metal is hot, or in thin flat stock, where there is not enough heat to work, the light fin can be ground off.

Welding Machines

The electric current transformer, the clamping device and the pressure device are the three necessary elements in an electric welding machine. Although it is possible to give the transformer a different location from that of the two mechanical elements, it is rarely done, the commercial welding machine employing all three elements in the same structure. There are, of course, many special departures from this general form. The mechanical and electrical controls may be operated by hand, by foot, or may be automatic and operated by power. For small work, like welding wire, spring pressure in forming the weld is usually employed, and clamping is done either by hand or by power; for metals like copper and brass a weight pressure is usually best; for rounds and like sections up to $\frac{3}{4}$ inch, hand pressures are usually employed; for larger sections, hydraulic pressure or pressure obtained through self-contained oil jacks is used.

Electric welding machines are necessarily more or less special in the construction of their clamps and electrodes, no one machine being suitable for a great variety of sizes or forms of work, and some are entirely special and suitable only for the work for which they are built. Some are called semi-automatic, as when the operator's duty is only the putting in and taking out of the pieces, while other machines are entirely automatic, as regards clamping, exerting pressure for welding, and controlling the current.

Power machines for spot welding are built with heating time-adjustments and regulation so that the machine can be set for the right speed and the correct time of heating required for varying thicknesses and conditions of stock. They can be operated continuously or intermittently as necessary, so that the greatest amount of work which the operator can handle will be taken care of by the machine.

Time and Current Required

There is no process by which heat can be delivered to metal so quickly as by the electric resistance method. In small wire a fraction of a second only is required for welding, while with larger pieces proportionately more time is required. Welds can be made either quickly or slowly, depending upon the amount of power available. It is always desirable to have ample power so as to make the weld in the quickest possible time, as better results are usually obtained, and time and power are saved. A ¾-inch round can be welded with 15 KW. in 15 seconds or with 23 KW. in 6 seconds. Endless pieces like rings take more power, as the diameter decreases; copper and brass require more power and less time than steel or iron of like section.

The expense for current is small, as it is used only while the pieces are heating in the welding machine, which is from one-quarter to perhaps one-half of the time of the day's run. The welding machine is always ready if the current is available, by simply closing a switch, and when the weld is made the expense of consumption of energy instantly stops. The output per day for any welder depends, of course, upon the size of the stock to be welded, the shape of the pieces, and the facilities for handling. In wire hoops, under the best conditions, 1000 welds can be made per hour, while in very heavy tires, 100 per hour would be a very good output.

Large volumes of current at low pressures are required, approximately from 2000 to 50,000 amperes, at from 1 to 7 volts, being used for welding from ¼ inch to 3 inch round, or equivalent section. The pieces to be welded, when mounted on the terminals of the transformer, complete an electric circuit, all parts of which, except the pieces to be welded, are adequate to carry this heavy current without heating excessively or without much loss of energy. The current depends on the section of the pieces and the speed of welding.

An alternating current wired from a single phase, or from one phase of a multi-phase system is the most convenient and is universally used. A current of any usual frequency can be employed, the welding tranformer being adapted for the conditions, 60 cycles generally being used, although welding machines are employed on circuits of 25 cycles and even on circuits as low as 11½ and 15 cycles. The pressure also may be any commercial voltage, 220 volts being commonly available, although welding machines are operated on circuits ranging from 104 volts to 550 volts.

This current is usually obtained from the local lighting and power company rather than from individual generating plants, although in cases of large electric welding installations individual generator equipments are used. In installations of single welding machines, the current being used intermittently, the actual kilowatt-hours consumed in a day's run are small and the expense of current so light that an individual generator is not desirable.

The welding machine being a machine with self-induction, the power factor is low, sometimes 50 per cent, varying according to conditions, but rarely higher than 70 per cent. The nature of the load on the generator supplying the current is different from that of almost any other current consumer, except those using some forms of alternating-current motors, on account of the fact that when the circuit is closed through the breakswitch, the full amount of power required to make the weld is instantly thrown on the generator; and instead of building up from a minimum to a maximum, the maximum is first demanded of the generator. This necessarily creates some disturbance on the line.

Strength of Welds

When the visible surface of a piece (outside) reaches a welding heat the interior is necessarily also at a welding heat; for this reason, a more uniform result and a greater percentage of good welds can be obtained, and at least as strong, if not a stronger weld can be made than by any other method. In general it may be said that in good open-hearth mild steel, almost, and in some cases, fully, the strength of the metal section is obtained. Bessemer steel, being frequently higher in sulphur and phosphorus, cannot always be welded with such good results.

In high carbon steel the heat is so distinctly localized and its dissipation is so fast when the current is withdrawn, that just at either side of the weld the metal is chilled and may be more or less brittle. To overcome this, it is the practice to reclamp with the electrodes wider apart, letting the current flow through the brittle parts until they come to an annealing temperature. Reheating may be repeated when needed. A spot weld frequently is much stronger than a rivet. In very thin sheets the metal pulls away leaving a hole in one sheet with the metal of the weld adhering to the other sheet.

There is an infinite variety of welding work which can be done by the electric resistance process, an interesting example of which is the welding of rails for street railways, taking the direct current from the overhead trolley, transforming it to alternating current, and welding plates across the joint, making a practically continuous rail.

This work is being very extensively done in the large cities. An unusual method of making agricultural wheels consists in assembling the spokes in the two halves of the hub and mashing them all together at one operation in a welding machine. The welding of wire fabric used in wire fencing and for concrete reinforcing, for which has been developed the only fully automatic electric welding machine in existence, should also be mentioned; the strand wires and the stay wire are fed into the machine and welded automatically at all intersections.

A more detailed review of the conditions relating to electric resistance welding was presented by Mr. A. E. Buchenberg in Machinery. In contemplating the practicability of using electric welding machines, says this writer, the principal questions that should be given serious consideration and be definitely determined by the manufacturer, are as follows: 1. The efficiency and reliability of electric welds. 2. The output of machines in welds per hour. 3. Adaptability of electric welding machines to his work and shop requirements. 4. The cost of operation. 5. The initial cost of machines and such auxiliary apparatus as may be required.

The following pages will be devoted to a general discussion of electric butt welding as opposed to brazing and the ordinary forge method of lap welding, with particular reference to shop requirements.

Efficiency and Reliability of Welds

The manufacturer is not so much concerned with the fact that perfect welds can be made electrically, but more vitally interested in the efficiency, uniformity, and cost of the welds, that he may reasonably expect on his product and under his shop conditions. In some classes of work there can be no allowance made for even a very small per cent of breakage from imperfect work, and every weld must be a perfect molecular union over the entire area of the welding surfaces. As an example, we may take the case of the steering mechanism of an automobile, where it is found economical in machine work and stock to electrically weld the threaded steering head to the tubular stem. Under service conditions, this weld may at any instant be called upon to withstand severe longitudinal and torsional stresses which can very nearly reach the ultimate safe strength of the tube's cross-section. The safety of the occupants of the car may depend upon the efficiency of this weld, and before adopting the electric process, the manufacturer must be convinced that the welds can be made under commercial conditions so that each and every one can be absolutely depended upon.

There are many instances where the electric weld, if reliable and practical, will reduce production costs very materially in eliminating expensive machine work and the present unavoidable waste of stock. This is especially true where a great reduction in diameter is required over a considerable length of a bar or rod. It then becomes convenient to weld a rod of one diameter to another of greater diameter which results in a saving both of stock and the expensive machine work which would otherwise be necessary to remove it. Where a clevis or an eye is required on one or both ends of a rod, drop forgings might be welded to a length of cold-rolled steel. A machine bolt in place of being turned from hexagon stock might be made of two pieces, the head—an automatic screw machine product—welded to round stock of the proper bolt diameter. Where large or complicated drop forgings are required and the initial cost and upkeep of the dies would be high, the part might be made in two or more small drop forgings welded together, and allow the use of simple and comparatively inexpensive dies. From the few examples given, it will be plain that the question of reliability of electric welds may determine to a very great extent the shop production costs, assuming of course that the expense of making the welds is low.

The quality of any weld, whether made by the blow torch method, the ordinary forge method—usually called a "fire weld"—or electrically, depends entirely upon the efficiency of the molecular union between the welding surfaces. With either the electric or the fire weld, the molecular attraction, or cohesion, is brought about by first heating the stock to a plastic semi-fluid condition and then forcing an intimate surface contact between the two pieces by a succession of blows, as in the ordinary fire weld, or by the application of a heavy mechanical pressure as in the electric process. The "scarf" or lap of the fire weld is a convenience for the application of the blows of the hammer while making the weld and in many cases is a requirement, as when welding surfaces equivalent in area to at least the cross-section of the stock. With the electric process, no scarf or other preparation of the stock is required, the two pieces to be welded being simply clamped in suitable jaws or dies with their ends abutting, the welding pressure then being applied axially.

The Electric Welding Machine

The electric butt-welding machine which in some of its highly-developed special and automatic forms may be a very complicated piece of mechanical and electrical apparatus, is a structure for first heating stock by means of an electric current and then exerting mechanical pressure to force the welding surfaces together.

The component parts of an electric butt-welding machine in its simplest form are as follows:

1. A special type of transformer whose primary coils are connected to an alternating current-supply circuit and whose secondary winding delivers an output of very low voltage but heavy current. The transformer may be operated from any alternating current single-phase circuit of standard voltage and commercial frequency. The usual lighting and power voltages are 110, 220, and 440 volts, while the frequency may be either 25 or 60 cycles. If necessary, the welding machine may be operated from a 133 cycle circuit. Where polyphase alternating current is used, the welding transformer can be connected across one phase of a two- or three-phase circuit. Fig. 1 shows in detail the construction of a typical welding transformer with the

connecting leads to the clamping dies attached to the secondary winding. On account of the low voltage required, the secondary winding in this instance takes the form of a solid copper casting extending through the laminated iron core. It will be noted that the secondary leads are each made up of a large number of thin copper strips to afford the necessary flexibility for motion of the clamping dies.

2. Two copper clamping dies and supports in which the stock to be welded is securely held to afford good electrical contact and to prevent shifting and displacement of the work under end pressure.

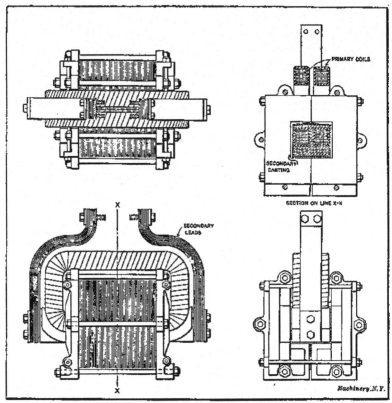

Fig. 1. Typical Transformer for Welding Machines

The dies and supports are capable of a limited movement toward and away from each other in suitable guides. In the machine as ordinarily constructed, the left-hand die is stationary but capable of adjustment, while the right-hand support is movable and connected to the compression mechanism. Each die and support is connected to one of the flexible secondary leads of the welding transformer.

3. To afford the heavy mechanical pressure necessary to be exerted at the proper time to force the heated abutting ends of the stock together to form the weld, a number of different arrangements are made use of. The compression mechanism used on a particular ma-

chine will depend to a great extent upon the size of the stock to be welded. For the smaller work, a spring or simple toggle lever is used, and for heavier stock, gears operated by a pilot wheel or a hand-operated double-acting hydraulic jack.

Machines for heavy work are seldom made automatic in their operation, since the question of large output per hour is not so important, and it is not always possible to supplant human judgment and skill with mechanical automatic devices.

Fig. 2 shows a simple form of the Toledo Electric Welding Co.'s machines for welding straight rods or tubes. The clamping dies A are fitted to the work they are to hold, and are mounted on the sliding supports B and B_1. The supports are mounted on guides shown at C. The clamping dies are operated to grip the stock D by means of the clamping levers E. The left-hand head B is stationary while welds are being made, but it can be adjusted for position by means of the shoulder-screw F. The compression toggle lever G is connected to the right-hand head B_1, by links as shown at H. The welding transformer J can be seen through the opening in the side plate of the machine. The foot switch for closing and opening the current through the primary coils of the transformer is shown at K.

Operation of an Electric Welding Machine

The several steps in the operation of the machine when making a butt weld are as follows. Two pieces of stock are clamped in the dies with the surfaces to be welded opposed and abutting, the dies being separated from each other a short distance to allow a converging motion for compressing the stock at the proper time. A switch connecting the primary coils of the welding transformer to the supply circuit, and which may be hand- or foot-operated, as convenience may dictate, is closed. The induced secondary current of the transformer now flows through the heavy flexible connecting leads, through the clamping supports and dies into the stock to be welded, and across the abutting surfaces. The junction of the welding stock is the point of highest electrical resistance in the entire transformer secondary circuit, which is made up of the secondary winding, connecting leads, clamping supports and dies, and the small projection of stock over each clamping die. The design of the transformer, secondary leads, clamping supports, dies, etc., makes their combined resistance very small as compared to the contact resistance at the point of weld. In conformity to the laws governing the heating of conductors carrying electric currents, practically all the heating will be confined to this point. In other words, nearly all the electrical energy taken from the supply circuit will be concentrated in this one location in the form of heat.

The secondary voltage of the transformer is so designed that the volume of secondary current forced through the junction of the two pieces of stock will produce a welding temperature at this point in a certain predetermined time. The actual secondary voltage required will depend upon the cross-section of the material to be welded, and whether the stock is iron, steel, brass, copper, or aluminum. The voltage varies between one and seven volts.

A voltage regulator of the inductive or "choking" type is usually supplied with each welding machine. This regulator is an auxiliary piece of apparatus connected in circuit with the transformer primary coils, and by means of which the secondary voltage can be readily adjusted through a wide range to afford the best operating conditions on varying kinds and sizes of stock.

At the instant a welding temperature has been reached, the switch is opened and the stock quickly compressed under heavy pressure to form the weld. A small amount of semi-fluid material is displaced under the pressure and thrown out all around the stock at the point

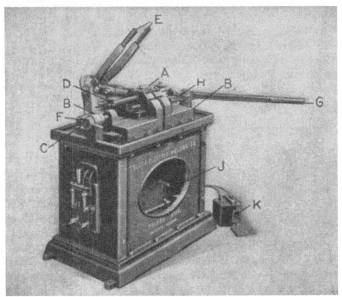

Fig. 2. Simple Form of Machine for Welding Straight Rods or Tubes

of weld in the form of a fin or burr. When necessary, this surplus metal can be removed by grinding or chipping, or it can be reduced under a power press to the stock dimensions.

Conditions Necessary for Perfect Weld

The primary conditions necessary to make a perfect weld between similar or dissimilar weldable metals are as follows:

1. The welding surfaces must be clean.

2. Each of the two pieces to be united must be at its particular welding temperature. The entire surfaces to be welded must be at this temperature, or in other words, the heat distribution must be uniform.

3. Repeated blows or a heavy continuous pressure must be applied while the welding surfaces are each at the proper heat, in order to form an intimate union between the two pieces of stock.

In the following are taken up in detail some of the more important conditions as they exist in the operation of an electric welding machine and in the fire method of lap welding:

Conditions of Welding Surfaces

The primary requisite, that the welding surfaces of the stock must be clean, is fully met in the electric process. Furthermore, the abutting welding surfaces are practically excluded from the air while being heated, and with the short time required to bring up the temperature, little or no oxidation can take place; for this reason, no flux of any description is required even on brass, copper, and aluminum. With the fire process, the welding surfaces are exposed to the action of impurities, particularly surphur in the coal, or the products of combustion in an oil or gas flame. Under these conditions and the length of time required to heat the stock, the use of a flux as a protective covering against oxidation over the welding surfaces, becomes an absolute necessity.

Heating

In the case of the electric weld, the heating begins in the interior of the stock and travels out toward the surfaces so that every particle of metal at the point of weld is at a uniform temperature. This condition is automatically attained, since the flow of current will always be greatest through the path of least resistance. If, on account of varying surface contact resistance, one part of the stock should heat up more rapidly than another, the increased resistance due to the higher temperature would automatically shunt a greater portion of the total current through the cooler part of the stock which is of lower resistance. This action, in combination with heat conduction, would result in an even temperature throughout the stock at the point of weld.

The heating action is concentrated at the junction of the two pieces to be welded, as the time of current flow is so short that the heat travels back but a short distance each side of the weld by conduction. There is no scaling or pitting due to surface oxidation, and the heat discoloration of the material in the case of round stock is seldom visible on each side of the weld for a distance greater than the diameter of the stock. All the heat is concentrated where needed, and there is no waste of energy or fuel in the useless heating of a considerable length of stock on each side of the weld.

The work is always in plain view of the operator who is able to judge to a nicety the instant at which the proper welding temperature for any particular grade of stock is reached. This is an important factor in obtaining perfect welds between materials of widely varying chemical and physical properties, where the proper welding temperature for each material may be at wide variance. Specific instances are the welding of cold-rolled steel rods to drop forgings or the welding of steel stems to brass bolt heads.

With the fire weld the heat is, of course, applied to the surface of the stock and the interior is heated by conduction only. With an intense fire and under conditions of rapid shop production, the outer surface, and especially the thinner edges of irregular sections, may easily be at a higher temperature than the heavier section. This condition may, and unfortunately often does, result in imperfect welds. It is particularly noticeable on lap welds.

With a fire weld using oil, gas or coal as fuel, a considerable length of the stock is brought to a high temperature. There is always more or less scaling and pitting of the stock owing to the length of time required for heating, during which time the surface of the stock is exposed to the oxidizing action of the air. It is a practical impossibility to forge-weld brass and other alloys of copper as the component metals of low fusing point will volatilize before the copper has reached a welding temperature. The stock is buried under a cover of coal or partly hidden in the flames of a gas or oil fire so that it is difficult to judge the temperature without uncovering the stock or removing it from the fire. The result is that the stock is, in many cases, underheated or overheated, the consequence in either case being an imperfect weld. When the output of welds per day is large, a very considerable saving of stock is effected by using electric welding machines, since the amount of stock wasted in the upset or fin is much less than the stock required for the overlap and scarf for a fire weld.

There is no danger from an electric shock to the operator of the machine since the primary coils of the transformer are heavily insulated, and the possible voltage to which the operator is subjected is no more than that of the ordinary door bell battery, and so slight that it cannot be felt under any conditions.

Output of Machines

The output, say in welds per hour, of any machine, will be determined by both the electrical and mechanical design. With automatic machines designed for a particular piece of work on light stock, the output is large. As an example, a machine for welding wire barrel hoops will take the wire from the reel, cut it into the proper lengths, and deliver the welded hoops at a rate of approximately 650 per hour. In the case of a hand-operated machine, the output will be determined by the mechanical design of the clamping dies and compression mechanism, the time required to heat the stock, and the facility with which the stock can be inserted in and removed from the machine as determined by its general shape and welding cross-section. Fig. 3 illustrates a hand-operated machine whose output on straight stock with a welding area equivalent to the cross-section of ¾-inch round stock, will be approximately 250 welds per hour. As a general rule, the larger the stock, the smaller will be the machine output, both on account of the longer time required for heating and the greater length of time required to handle the heavier stock.

Adaptability of Electric Welding Machines

Except in the case of a welder especially designed for one particular piece of work, quite a range in the shape and size of stock can be handled by one machine. A welder equipped with a voltage regulator can be adjusted to weld stock much smaller in sectional area than the rated capacity. As an example, a welder whose maximum capacity is one inch round stock, or an equivalent cross-sectional area in an irregular section, will, with a proper adjustment of the regulator, weld one-quarter inch round stock. However, as a commercial proposition, it is not good practice to weld very small stock on a large machine, as all the working parts are necessarily heavy and cumbersome on

Fig. 3. Hand-operated Machine with Output of 250 Welds on Straight Stock with Welding Area Equivalent to ¾-inch Round Cross-section

light work. For this reason the output would naturally be less than with a smaller, lighter, and more easily operated machine. Usually a change in the size of stock to be welded occupies only a few moments' time to change the clamping dies to conform to the new stock, and make the proper adjustment of the voltage regulator.

From the standpoint of maximum output, an important consideration in the selection of a welding machine is the facility with which the stock can be placed in the clamping dies, and the proper arrangement of jigs for accurate alignment of the work. Welders are now designed in standard forms for different general classes of work, and a machine whose welding capacity is ample for a particular piece of work, might be entirely unsuitable for economical production on account of the mechanical design. For instance, a machine designed for welding straight bars or rods would be impractical for taking care of such work as vehicle dash frames; these require a special machine designed so that the frame may be swung into several positions.

Cost of Operation

The operating cost will depend upon the size and material of the stock to be welded, upon the cost of the current, and the number of welds made. No current is used while the stock is being inserted into the clamping dies of the machine, or while being removed after the weld is completed. The amount of electrical energy required will depend upon the kind and shape of the material, and its cross-sectional area. The actual cost of operation is very low as is indicated in the following table, which gives the time and kilowatts per weld required for a number of sizes of iron or steel stock. The tabulated cost per 1000 welds is based upon a unit current cost of one cent per kilowatt hour. The actual cost in any particular instance can be determined by multiplying the cost per 1000 welds as given in the last column of the table by the price of current per kilowatt hour at that locality. The costs given do not include the time of the operator.

Initial Cost of Machines

The first cost of a welding machine will depend to a great extent upon the sectional area of the stock to be handled, and whether the material is iron, steel, brass, aluminum, or copper. The higher the electrical conductivity of the metal, the greater the amount of current required to raise it to a welding temperature in a given time, and the larger the welding transformer required. The cost of the machine will also be governed to some extent by the shape of the section of the stock quite independently of the actual sectional area. Heavier and more expensive clamping dies will be required to weld stock $\frac{1}{8} \times 6$ inches than would be required for the same area of metal in the form of round stock. In the case just given, special mechanism must be used in connection with the clamping jaws in order to obtain an equal distribution of current along the abutting edges of the stock. Where a great output in the number of welds per hour is demanded, automatic or semi-automatic features become necessary and the cost of the machine is materially increased. Machines built for special work or to meet extraordinary conditions are, of course, much more expensive than standard stock machines. Up to sectional areas equivalent to $\frac{3}{4}$-inch round stock, machines are usually operated by means of a simple hand toggle lever. From $\frac{3}{4}$-inch to 1-inch round, a handwheel operating through gears may be used. For larger stock it becomes necessary to resort to a special double-acting hydraulic jack.

From the foregoing it will be seen that the first cost will be governed by size and kind of stock, shape of the parts to be welded, and the capacity of the machine in welds per hour.

While a welding machine, especially the smaller sizes, can be connected directly to the circuit supplying light or power to the shop, it is usually better, on account of the line disturbances set up by the intermittent inductive load, to install a separate transformer to supply the welder only. This transformer is usually furnished by the local power and lighting company. Where alternating current is not available, a small alternator driven from the line shafting can be installed to operate the welder.

Welding of Dissimilar Metals

A valuable feature of the resistance process of electric welding is that dissimilar metals can be perfectly joined. This possibility permits combination of metals best suited to the conditions of use to be made, as well as very substantial economies in the use of high-priced materials.

As examples of what can be done in the welding of dissimilar metals may be mentioned a number of products regularly made for the

TABLE GIVING TIME AND COST OF WELDING VARIOUS SECTIONS IN IRON OR STEEL

Diameter, inches	Area in square inches	Kilowatts, Transformer	Seconds to make Weld	Cost per 1000 Welds, Current one cent per Kilowatt hour
¼	0.05	5	5	$ 0.07
⅜	0.11	7½	6	0.13
½	0.20	8	10	0.22
⅝	0.31	10	12	0.33
¾	0.44	12	15	0.50
⅞	0.60	15	20	0.83
1	0.79	18	30	1.50
1⅛	0.99	20	30	1.66
1¼	1.23	26	40	2.89
1½	1.77	40	60	6.67
1¾	2.41	45	70	8.75
2	3.14	56	80	12.44

market. Poppet exhaust valves for high-speed gas engines are made with a carbon-steel stem electrically welded to a nickel-steel head. By making the head of nickel steel, or some other alloy steel suited for the purpose, the very best metal is put into the head of the valve, which part is subjected to the hardest usage. Nickel steel is peculiarly suited to the trying conditions surrounding gas engine exhaust valves, because it does not pit, warp or corrode as does common steel in such a situation. It is also much tougher and is not apt to break because of the hammering it receives. On the other hand, the stem made of carbon steel stands the wear in the guide better than would a nickel-steel stem. It is also stiffer and can be hardened on the end. This latter condition is of considerable importance, because a high-nickel steel cannot be properly hardened to withstand the hard blows of the valve mechanism. Another important advantage of the combination is the saving of the high-priced metal and the obvious saving of labor over that required for making and machining a forging.

Cap-screws are made with brass heads and steel bodies. Such screws are fifty per cent stronger than would be screws made entirely from brass. This combination is made for use in places where an ornamental finish is required, but where the head only shows and a brass body is of no advantage.

Other valuable combinations of dissimilar metals could be mentioned, but the foregoing will serve to illustrate the advantage of electric welding in this line. One claim made for electric welded all-steel bolts and screws is that they are stronger than when made by the ordinary methods. The reason is that the die-drawn surface of the stock is retained on the body, this portion of the body being much stronger than the center which is left when the bar is turned down to the body size from the head size. This fact is indicated by tests made of electric welded cap-screws and cap-screws made by twelve makers, from ordinary stock. The tests show that the average tensile strength of ½ to 1¼ inch electric welded cap-screws was 97,862 pounds per square inch, while the average of the ordinary stock screws was only 56,570 pounds per square inch. The difference in favor of the electrically welded screw is thus 73 per cent.

Electric Welding of Copper, Brass and Aluminum

The welding of brass, bronze and other alloys of copper is almost impossible as a forging operation. The fusing points of the several alloy metals are considerably below that of copper, and it becomes a very difficult matter to prevent the oxidation of these metals before the copper component has reached a welding temperature. While it is possible to weld copper and aluminum in the same manner as iron or steel by the forging method, the work is more or less difficult and requires a careful and skilled operator. Some of the difficulties in welding are as follows: While the metal is at or near the welding heat and exposed to the air, a very rapid surface oxidation takes place, and the oxide or scale formed is extremely difficult to treat with any flux. The range of temperature between the heated plastic or welding condition and the fusing point of the material is very small. To add to the difficulties the metal becomes brittle as the temperature approaches the welding heat.

While aluminum reaches a welding heat at a temperature considerably below that of copper, it is also subject to a serious surface oxidation when exposed to the air at high temperatures. The range of temperature between the welding and fusing points of aluminum is only about 180 degrees F. If overheated it will simply spatter away under the hammer when attempt to make a weld is made.

The principle for the electric welding of copper, brass and aluminum is the same as for iron or steel, as already explained, *viz.*, forcing the two welding surfaces into intimate contact while each is in a heated plastic condition.

The time required to electrically weld copper, aluminum and copper alloys depends to a very great extent upon the cross-section of the stock, and varies from one second or less on very small stock to a minute or more on the larger sizes. The time limits between which stock of any given cross-section can be successfully welded are comparatively wide and governed in both directions by the volume of the heating current through the point of weld. If the current is low, the temperature rises more slowly at the point of the weld, and the heat travels back a considerable distance on each piece by conduction before the surfaces to be welded reach the required temperature. Under these conditions the fin or upset becomes quite large and entails too much expense in grinding to remove it. In the extreme case of insufficient current the heat is carried away from the point of the weld by conduction to the copper dies and by direct radiation to the air so rapidly that a welding temperature cannot be attained. If the volume of current is too great a very rapid heating of the stock takes place and trouble is experienced due to the fusing and oxidation at the point of weld which occurs more rapidly than the pressure can follow up the softening of the metal, and the excessively heated stock "spatters." Although the heating current is automatically cut off at the instant the forward motion of the die begins, the oxide coating on the end surfaces prevents a molecular union and a perfect weld. In other words, the stock is burned.

The heating current must be adjusted between the two extremes given above to a point where it is intense enough to bring the stock to a welding temperature, yet not so great as to cause an excessive heating and a "blowing out" of the weld. The correct adjustment is not a difficult matter, and is attained by varying the voltage impressed upon the primary coils of the transformer in one of two ways.

If the electric welder is operated by an alternator carrying this machine only, the voltage of the alternator (which in this case is the same as the primary voltage of the transformer) can be varied by a manipulation of the alternator rheostats.

Again, when the electric welder is operated from a circuit the voltage of which must be maintained at a constant value, such as a power circuit from which other welders or motors are being operated, an inductive regulator is used as an auxiliary apparatus to the welding machine. The action of such a regulator is simply that of a variable choke coil, and any desired voltage can be obtained across the primary terminals of each welding transformer by a proper adjustment of the regulator.

A skilled or experienced man is not required for operating an electric welding machine, as the only duties of the operator consist in clamping the work in the dies of the machine and closing the switch. For this reason it is customary to use boys on this work. Many machines for light work are now made automatic in their operation, requiring no attention beyond feeding in the stock to be welded.

Uniform and perfect molecular union is obtained with this process, since the heating of the stock is from within outward and the entire areas of the welding surfaces are at the same temperature. The strength of the weld is practically equivalent to the strength of any other section of the stock of equal area and will withstand any subsequent bending, rolling, hammering, or drawing process to which it may be subjected.

The difficulties encountered in the forging process due to the oxide surface films formed at high temperatures are not present in the electric process. During the extremely short heating period the welding surfaces are in contact and practically excluded from the air. Furthermore, the heating action ceases the instant the welding

TIME, CURRENT CONSUMPTION AND UNIT PRICE FOR ELECTRIC
WELDING OF COPPER

Area in square inch	Approximate Equivalent diameter, inch	Kilowatts	Time in Seconds to make Weld	Cost per 1000 Welds at 1 cent per KW. hour, Dollars
0.05	¼	5	4	0.055
0.10	⅜	7	5	0.097
0.20	½	14	7	0.272
0.30	⅝	19.5	9	0.487
0.40	11/16	26	12	0.866
0.50	13/16	32	14	1.240
0.75	1	46	17	2.170
1.00	1¼	62	21	3.610

temperature has been reached, and the heat extends to but a very small distance on each side of the weld. It is obvious that with a continuously applied pressure which instantly compresses the stock to form a weld when the proper temperature has been reached, the difficulty experienced in other processes due to the small range of temperature from the plastic to the fused state of the metals is overcome.

The electric welding of copper, brass and aluminum is a very rapid operation and entirely free from noise, dirt and smoke. The machine can be located in any convenient position in the shop and is free from danger of electrical shocks to the operator. The motion of the movable die can be adjusted for both the forward and backward travel so that all welds are to gage. This is an important consideration when, for example, many thousands of rings must each be welded to an exact diameter.

The cost of electric welding is low as compared to other methods. Herewith is given a table for copper showing the kilowatts and time required to make a weld from the time of closing the switch. Also the cost per 1,000 welds at a unit basis of current cost of one cent per kilowatt-hour. To arrive at the actual cost of the current per 1,000 welds it is only necessary to multiply the cost given in the table by the price for current per kilowatt-hour in any given locality.

CHAPTER III

POINT AND RIDGE METHOD OF ELECTRIC WELDING

The electric welding process, which from a commercial standpoint is comparatively new, has revolutionized many manufacturing methods, owing to its efficiency, particularly on that class of work which must be produced in large quantities and at a minimum cost. The extent, however, to which this method of welding metals is now employed, is no doubt very limited, in comparison with the number of manufacturing operations which could advantageously be performed by the electric process of welding. This is largely due to the fact that the art is little understood, particularly as regards the variety or range of work which can be efficiently welded electrically.

Various types of electric welding machines have been placed on the market by the Universal Electric Welding Co. One of the ma-

Fig. 1. Examples showing Possibilities of Welding Thick Materials

chines handled by this company is made to be fastened on a bench and operated by a hand lever. This same machine can be fitted to a stand and operated either by hand or by foot power, and can also be belt-driven and made to work automatically. It is particularly adapted for "electrode" welding, commonly called "spot welding." It is constructed with long projecting horns made as deep as three feet and with varying capacities to weld from the lightest gages up to ¼-inch sheets. The generally accepted idea that spot welding is practical to only as high as ⅛-inch thickness of sheet is fully disproved in the case of this machine. Pipe from 6 inches in diameter and up to 6 feet long can also be welded in machines of this type.

Methods of Welding

For a number of years electric welding was confined to the butt welding of rods, tubes, etc., and later the "spot" welding of flat stock by means of isolated welds of a limited area was developed. The spot method opened up an entirely new field, which, as the result of still later improvements, has been extended until, at the present time, the electric welding method of uniting metals is adapted to a great

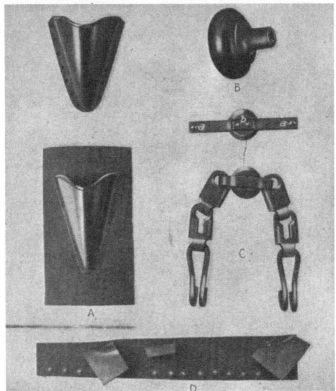

Fig. 2. Illustration of the Application of the Point Method of Electric Welding

variety of manufacturing operations. One method of making a "spot" weld is to use pointed copper dies or electrodes, which are brought into contact with the work which is welded by the passage of a large volume of low-voltage current. Another method is to raise projections above the plane surface of the parts to be welded, which serve to concentrate or localize the current in order to heat the metal to the welding point; and a still further development consists in raising ridges above the plane surfaces, which, when crossed by corresponding ridges, give the same practical results as the raised points, with the additional advantage of ease in assembling the parts prior to welding. This idea of using points or ridges in connection with electric welding has made it possible to weld, commercially, an almost unlimited variety of work, as is indicated by the accompanying illustrations. In the spot electrode welding of light-gage metals, a slight indentation is usually left in the surface at the weld, but by a method

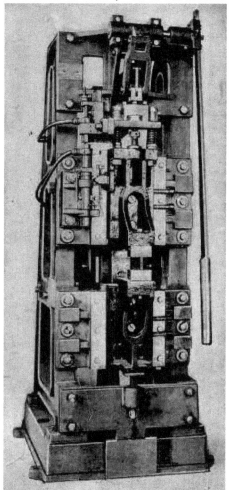

Fig. 3. Type of Electric Welding Machine for Heavy Work

recently developed, this condition is overcome, and the sheet can be left the full thickness and finished off. By using this method there is practically no limit to the thickness which can be spot welded. This is indicated in Fig. 1, where two bars each ⅞ inch thick have been spot welded together. The operation took about one minute.

Point or Projection Welding

As mentioned, the welding of sheet metal is not restricted to one spot at a time, for any reasonable number of welds can be made at one operation by the method known as "point" or "projection" welding. In such welding, used, for example, for cooking utensils, sash pulleys, etc., the parts when stamped have small projections raised above the plane surface of the metal, the height of the projection varying according to the gage of the material. This is done during the operation of stamping. When welding such parts, properly shaped copper electrodes are fitted to them. Each point acts as a resistance to the passage of the welding current. The current divides itself among these points and by their resistance to its passage, each becomes a heated welding point. Pressure applied to the softening metal completes all the welds simultaneously. Fig. 2 shows a spout welded to a coffee pot at twenty-three distinct points.

A marked distinction exists between the two methods of "spot" and "point" welding, as many cases occur when the spot method cannot be used, but the point method proves perfectly successful and commercial. Fig. 2 illustrates an anti-skid chain which is an excellent example. In making this chain the electrode method was first tried by the manufacturer. The point method was then applied, raising points on both the body and the strap, and resulting

Fig. 4. Automatic Electric Sheave Welder

in a perfect weld. The output by the spot method was 600 per day of unsatisfactory welds; by the point method, 3000 good welds.

Another excellent example is that of the door knob shown in Fig. 2. The shank is welded to the hollow knob by six distinct points. It is impossible to weld this except by the point method.

Ridge Welding

To facilitate the assembling of parts, a further improvement was made by the introduction of so-called "ridge" welding. The result of two ridges crossing is the same as of two points, and the operation of making the weld is identically the same. Both point and ridge welding permit the use of large flat blocks of copper for electrodes, the heat being localized by the points or ridges forming the welding spot. In both these methods the electrodes require very little attention except an occasional touch with a file over the surface, as against continual shaping of small pointed electrodes.

Types of Welding Machines

The machine used for point and ridge welding is equipped with large copper electrodes instead of the pointed type, and the current is

Fig. 5. Sash Pulley and Housing welded by the Point Method

concentrated at the points to be welded by the small raised projections, as mentioned, instead of by reducing the area of the electrodes. With the large electrodes, sufficient current for welding heavy stock can be conducted without excessive heating and deterioration. A welding machine having these large electrodes is illustrated in Fig. 3, which shows a standard type intended for general work. This machine may be either hand-operated, as shown, or may be made semi-automatic, belt-driven. When a great quantity of similar articles is to be made, special machines are usually built. The automatic sheave welder shown in Fig. 4 is used by the American Pulley Co.; it welds 15,000 pulleys in ten hours.

Examples of Electric Welding

Fig. 5 shows a sash pulley and housing which is an example of point welding. The pulley is made in halves, one half having an annular ridge and the other, six projections or raised points A. When the points of one half are brought into contact with the ridge on the other, the electric current, being concentrated, fuses the metal instantly, and six homogeneous point or spot welds are produced. These pulleys are welded on the automatic machine shown in Fig. 4. The housing *B* is also made in two parts, each of which is joined to the base-plate by four point welds as shown.

Another example of point welding is shown at *A* in Fig. 2, indicating how a spout is welded to a coffee pot, as already referred to. The "anti-skid" chain for automobiles shown at *C*, and also previously mentioned, has the central link *l* welded as indicated. The link is drop-forged with raised points at *a* and *b* which form the welds after the wings are bent over onto the central part. The strength of the point

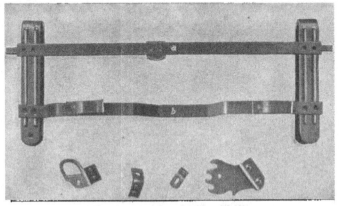

Fig. 6. Examples of Ridge Welding

weld is indicated by the sample shown at *D*. An attempt to tear the small pieces from the steel strip, resulted in shearing the metal around the weld, but in no case did the weld prove defective.

Fig. 6 shows an example of ridge welding. This is part of a go-cart frame. The stock from which the end pieces are made has ridges rolled in it as shown. The cross-bar *a* is provided with two raised points at each end that come directly over the ridges, and the bar *b* has concave ends thus giving contact points where the curved· edges rest on the ridges. Beneath this frame a number of small parts are shown that are prepared with projections ready for welding. The ridge method of welding is also used for welding reinforced concrete frames, the ridges which form the points of contact being originally rolled in the stock. Generally speaking, the ridge method is prefer-able, owing to the greater facility in assembling parts prior to weld-ing; the ridges also stiffen and strengthen the material. The results obtained by the ridge method are, as far as the quality of the weld is concerned, practically the same as when projections or points are used.

CHAPTER IV

ELECTRIC ARC WELDING

Electric arc welding as a means of uniting metals—particularly iron and steel—has been rapidly developed in the past few years, and apparatus for doing this work is now a standard product with a number of manufacturing concerns. This process of welding is particularly applicable to certain classes of work encountered in foundries, railroad shops, tank and boiler shops, steel mills, locomotive shops, and shipyards; and the demand for welding apparatus from these sources is well established. In addition to the field covered by these industries, where the use of this process has become more or less standardized, there are countless other lines of manufacture, each representing a great variety of work to which arc welding is adapted.

Various methods of using the arc for welding have been devised from time to time, the majority of which have met with indifferent success. At the present day practically all welding, in this country at least, is confined to the method in which an electrode and the object to be welded are connected in a simple electric circuit, and an arc of limited size is drawn between the two by bringing the electrode in contact with the work at the point at which the weld is to be made. The size of the arc is capable of adjustment to suit various classes and conditions of work.

The Carbon Arc

In practice there are two methods of applying this process to the making of welds and the cutting of metals. In the first, which makes use of the carbon arc, a rod of graphite forms the electrode; and the arc drawn between this rod and the work heats the latter to the point of fusion. This method is used for cutting or burning off metal, and is the simplest application of the arc. Its principal use is for reducing scrap material to sizes capable of being easily handled, and in foundries for cutting risers and fins from large castings. By extending this process of fusion and introducing pieces of metal within the influence of the arc, actual welding or building up of the work is accomplished. The metal supplied, which may be either in the form of small pieces of scrap material or a rod held in the operator's hand, is fused and unites with that part of the work already raised to a molten state by the heat of the arc, forming a solid mass of even structure upon cooling.

The principal field for the use of the carbon arc is in foundries and steel mills, for the repair of broken and imperfect castings of large size. The loss from this source, which is always high, can be reduced to a very small percentage, as castings containing blow-holes, cracks, shorts, etc., can readily be repaired with a small expenditure for material and labor. For all work of this nature in which the carbon arc is used, comparatively heavy currents are required, ranging from 300 to 600 amperes. Owing to the ability to use these heavy currents, and to apply the heat quickly and concentrate it at the required point, the heat generated at any particular point is very intense and the process of cutting or welding becomes a very rapid one.

The Metallic Arc

The second method in this process of welding makes use of a metallic electrode—usually of a soft grade of iron or steel—which during the operation of welding is fused by the heat of the arc and carried over in the form of small globules that are deposited at the point on the work from which the arc rises. The work itself is raised to a state of incandescence at this point, and the fused metal unites with it as it flows from the electrode. The operation of welding by this method is very rapid, as the fusing of the electrode is continuous after the arc is started, the drops of molten metal following each other in close succession. This method is extensively used in all classes of repair and reclamation work, such as filling in cracks of broken castings, building up the worn parts of rolls and rails, repairing cracks in boilers, patching locomotive fireboxes, and in many industries as a manufacturing means in the process of getting out the finished product. Examples of this latter use are the welding of heads and branches to tanks, joining the seams of tanks and boilers, welding fireboxes, flue sheets, boiler tubes, etc., and all classes of pipe and sheet metal work.

The current required for the metallic arc is small compared with that used in connection with the carbon arc, rarely exceeding 175 amperes for the heavier classes of work just described, and ranging from this down to as low as from 12 to 15 amperes for thin sheet metal work. The size of the electrode used also varies with the nature of the work and current required, the average being from 3/32 to ⅛ inch in diameter. That it is necessary in every case to have a proper relation between the current strength and the size of the electrode can be seen, when it is considered that the heat of the arc must be sufficient to raise a spot on the work to the point of fusion, in order that there may be actual union of the metal from the electrode with the work. If this condition of right temperature does not obtain, there will be an imperfect union of the oncoming metal with the work, and a poor weld will be the result. On the other hand, if the metal is overheated there is danger of burning it. Oxidation also takes place more rapidly, thus impairing the weld, and heavier heating and cooling strains are set up in the metal. The current must, therefore, be regulated to bring about the condition of a proper temperature rise in the work, and the size of the electrode should be selected to carry this current without danger of its being overheated and oxidized. On the other hand, the size of the electrode must not be too large for the current used, as this will result in slow and imperfect fusion, and equally slow and unsatisfactory welds.

Combined Use of Arc

In many cases, and more particularly in repair work, it frequently becomes necessary to remove parts of the metal at the place where the weld is to be made. For example, to widen out a crack in order that the metal from the electrode may be more readily deposited in it; or to cut out a burned, broken or worn spot for the insertion of new material. This operation of cutting is most readily performed by means of the carbon arc. In such work, therefore, the alternate use of the carbon and metallic arcs becomes desirable and to meet this requirement, as well as to make the outfit as general in its application as possible, means are usually provided whereby both classes of welding can be done from the same outfit. This feature also makes preheating possible, by which means work of large section is raised in temperature by use of the carbon arc, before the welding is actually done. The operation of welding on the hot metal results in the strains being more evenly distributed, both during the process of welding and when the work is cooling off. Welds of greater strength are thus obtained, and the structure of the metal in the weld is more homogeneous with that of the surrounding parts.

Description of Apparatus

The simplest possible outfit for welding would consist of a source of direct-current supply, an adjustable resistance for regulating the current, and an electrode holder. In practice, for reasons which will be explained later, the current is usually furnished by a low voltage generator which may be driven by a motor, engine or belt. In addition, the outfit usually includes a switchboard having on it the starting apparatus for the motor end of the outfit, if motor driven; the control and indicating apparatus for the generator, consisting of a field regulator, voltmeter, and ammeter; and the regulating apparatus for the arc circuit, consisting of a set of current regulating switches with resistance, and usually some form of automatic switch or contactor.

The generator should be compound wound in order that the voltage may be maintained constant under varying load. The need for close voltage regulation will be found to be greatest in connection with the metallic arc, and to increase as the size of the arc and the amount of current used decreases. The smaller arcs will be found to be very sensitive to even the slightest voltage variation, the direct result being an uneven deposit of metal, and burnt welds in the case of very light work. With the carbon arc, where the current used is generally large and where a certain amount of current regulation can be had by lengthening or shortening the arc, the need for close voltage regulation is not so great.

Of the resistance, a certain part is in circuit with the arc at all times when working, this resistance causing the difference between the voltage drop in the arc and the terminal voltage of the machine. It will vary with the amount of current required for welding, and is adjusted by the current regulating switch. When no contactor is employed in the arc circuit, the current at the time the arc is started is limited only by the resistance in that circuit, which is the amount required for welding. This may be of low value, particularly when using a heavy current. There is, therefore, danger of short-circuiting the generator until the arc is established and its resistance introduced into the circuit. The function of the contactor in the arc circuit is to cut out resistance after the arc is established, leaving in the circuit for welding that amount previously determined from the current to be used. By this means the chance for short-circuit is removed, and the apparatus made more automatic in its operation.

After the current is adjusted to give the size of arc needed, no further adjustment is necessary and the arc may be drawn and broken at will, the automatic character of the apparatus always insuring a return to normal conditions. By this means the operator is relieved of all concern as to current regulation, and his whole attention may be given to directing the arc over the work. The operation of welding by either of the methods described makes necessary the renewal of the electrode, though the rate at which the metallic electrode is consumed—owing to the fact that it constitutes the filling material—is much more rapid than that of the graphite rod. To facilitate the act of renewal or of feeding down as it is consumed, the rod forming the electrode is secured to a holder by some form of clamp that readily permits of its being released. The holder is designed to carry the current to the electrode with the least amount of heating of the operator's hand.

Owing to the intense nature of the light and heat rays from the arc, the necessity for careful protection of the operator's hands, face and eyes is very important. This is particularly so in the case of the carbon arc, where the volume of light and heat is very great. Heavy gloves serve to protect the hands, while for the face, some form of shield held in the hand or supported from the head is generally used. This is provided with an opening filled with several thicknesses of ruby or blue glass, which afford protection to the eyes and still permit of the welding operation being closely followed.

Potential Required for Welding

The potential which has been found to give the most satisfactory results for welding varies from 65 to 75 volts. A higher potential can, of course, be used, but as the drop in the arc rarely exceeds 65 volts, a potential in excess of this would have to be reduced by means of resistance in series with the arc. The wasteful effect of using a higher voltage, or of welding directly from shop or commercial circuits by means of resistance banks or water rheostats can be seen. The higher the voltage of the circuit from which welding is done, the greater the amount of resistance needed and the greater the energy loss due to this resistance. Assuming that 75 volts is the maximum required for all cases of ordinary welding, if a 220 volt circuit is used for this purpose, the efficiency is seen to be approximately 33 per cent, while at 500 volts it is as low as 15 per cent. It will also be found that when heavy currents—such as are required for welding—are taken directly from the line, serious voltage fluctuations will result, with corresponding ill effects on the apparatus connected to the line.

Flexibility of the System

Any number of operators may weld from the same outfit, each working independently of the other and taking the amount of current required for his own particular work, the self-regulating feature of the generator insuring a constant voltage. All of the arc circuits may be taken from the one welding panel or they may be divided among several smaller panels, which may be located at various centers at which it may be desired to do welding, these panels being connected by leads through the shop to the main panel. The latter, in this case, would contain only the motor and generator control apparatus. This arrangement is particularly desirable in locomotive and railroad shops, where the majority of the work is of such a nature that it cannot be moved around conveniently for welding. For doing work of this nature, the electrode holder is often fitted with leads of sufficient length to allow the work to be reached.

Welding can thus be done up to any distance from the outfit, the only limit being the allowable voltage drop in the lines to the work and the electrode. This, in turn, can be regulated to a certain extent by increasing the size of the cable as the distance increases. Beyond 500 or 600 feet, however, this method is hardly practicable for any work other than that which can be done with the metallic electrode, as the size of the cable required for carbon work with its large currents would increase to such an extent that its cost would be prohibitive and the handling of the cable exceedingly difficult. To meet conditions of this character a complete portable outfit consisting of generating and regulating apparatus, mounted on a truck that can be moved from place to place, is most appropriate. For land use the generator end of such an outfit is usually motor driven, while for marine work steam-driven outfits mounted on barges afford the most convenient arrangement.

Special Features

In connecting the work and the electrode in the welding circuit, the former should be connected to the positive side of the source of supply. There are two reasons for this, the first being that the positive side of the arc is by far the hotter of the two. The point on the work under the action of the arc is thus brought to the required fusing temperature in less time than if it were connected to the negative side of the circuit. A better distribution of heat between the electrode and the work is also secured by this means, as the electrode which is usually of small mass compared with the work should naturally be subjected to the less amount of heat. But a more important reason for this arrangement is that when the electrode is made positive the resulting arc is found to be very erratic and unstable, and its control becomes practically impossible.

It is not necessary that the operation of welding always take place
in a downward direction. While work with the carbon arc has to be
done in this position, due to the flowing of the metal in the weld, the
metallic arc can be used as readily on vertical or overhead welds as
on downward ones, the only difference being in the rate at which the
metal is applied. Owing to the fact that in any position other than
downward, the metal is applied against the force of gravity, its rate
of flow from the electrode is necessarily slower. This feature of being
able to weld with the work in any position occasions a great saving
in the amount of handling which would have to be done were it neces-

Fig. 1. Head, Flange and Branches welded in a 42-inch Tank

Fig. 2. A Built-up Fit on an Armature Shaft done with the
Metallic Arc

sary that all welding take place in a downward direction. The arc
process of welding is thus seen to be exceedingly flexible in its application, covering work of practically all classes and degrees of ac-
cessibility, and this feature greatly facilitates the operation of welding. Handling of the work is reduced to a minimum, and welds are
made with an ease and despatch not approached by any other method.

Character of Welds

A large measure of the success attained by this process is accounted
for by the satisfactory character of the welds from the standpoint of
efficiency. By a proper selection of the grade of filling metal, and the
exercise of care in making the weld, it is possible to obtain a tensile

strength in the weld of from 95 to 97 per cent of that of the original section. Welds made under the average conditions of everyday work show a tensile strength of from 80 to 90 per cent of the metal. It is possible by slightly reinforcing the welded section to make the strength of the weld even greater than that of the original section. This may be very desirable in many cases where a part has broken through having an undue strain put upon it. By a proper increase of section at this point, a repetition of the break may be avoided. Welds made by this process present a neat and finished appearance. With the metallic

Fig. 3. Fractured Mud-ring of Locomotive Firebox prepared for making the Welds

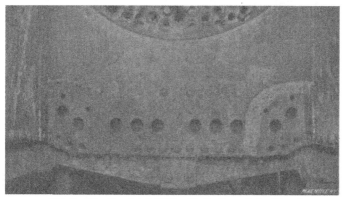

Fig. 4. The Completed Weld with Sections of the Throat Sheet replaced

arc, the filling metal added from the electrode can be deposited exactly where it is wanted; and with the carbon arc, where the added material is reduced to a molten state in the weld, it may be run at pleasure, extra material being added where needed and the surplus metal being fused down to the desired level.

Examples of Work Done by the Electric Arc

The illustrations in this chapter show examples of electric arc welding. While these are of a varied character and show work done with both the carbon and metallic arcs—with a considerable range of current—they do not in any way represent the complete possibilities of the process of arc welding and cutting. Fig. 1 shows an example of tank welding, in which the head, flange and branches of a tank 42 inches in diameter were welded with the metallic arc. The current required was approximately 165 amperes at 70 volts. The finished appearance of the welds and the necessity for little subsequent trimming will be evident from this illustration.

An armature shaft that had been turned too small at the spider fit is shown in Fig. 2; to remedy this error metal was added by means of the metallic arc, thus increasing the diameter sufficiently to provide

Fig. 5. Broken Casting from a Wood-working Machine

Fig. 6. Same Casting repaired with the Carbon Arc

for refinishing the fit to the required size. This was done with the metallic arc, using current of approximately 160 amperes. Figs. 3 and 4 show the repair of a fractured mud ring of a locomotive fire-box. It will be seen that part of the throat sheet has been cut away in Fig. 3 in order to give access to the mud ring. The fractures in the corners are first opened up with the carbon arc preparatory to welding, and after the weld is completed the sections of the throat sheet are replaced and welded as shown in Fig. 4. In this illustration, it will be noticed that the weld on the right-hand side has been dressed, while that on the left has not. The latter shows the appearance of the weld immediately after making a repair with the metallic arc. Figs. 5 and 6 show a broken casting of a wood planer before and after being repaired with the carbon arc. In cases of this kind the broken part is in use again in a short time, as the delay occasioned by having to replace it with a new casting is avoided.

Cost of Welding

The cost of making welds by this process can best be illustrated by examples covering operations of a common or familiar nature. The work capable of being done by arc welding is of such a varied character that it is not possible to give specific costs for each and every case that may present itself. The cost of generating current, the price paid for labor and the time required for doing any particular job will vary, and this will influence the cost of the weld. Of these three factors the first will be found to vary between the widest limits, the price of labor for the various classes of welding being fairly well standardized, and the time required for making welds not varying greatly when expert welders are employed. The cost of producing the following welds is figured on the basis of labor at 30 cents per hour, and current at 2 cents per kilowatt hour, the voltage in each case being approximately seventy.

A broken shaft 2 inches in diameter was welded and ready for refinishing in one hour; the current used was 350 amperes and the total cost 79 cents. A crack in the back sheet of a locomotive boiler 12 inches long was welded in nine hours, the current used was 175 amperes and the total cost $4.90. The risers on steel casting, 4 by 4 inches in size, were cut off in four minutes; the current used was 350 amperes and the total cost 5.2 cents. A cast-steel tender frame broken in three places was welded in twenty-seven hours; the current used was 300 amperes and the total cost $19.44. The journals of a worn 2-inch armature shaft were built up in three hours; the current used was 165 amperes and the cost $1.59. As an example of straight welding on sheet-metal work, seams on ⅛-inch steel can be welded at the rate of from 15 to 16 feet per hour, and on ¼-inch steel at the rate of from 12 to 13 feet per hour.

Any number of operators may work from the same outfit up to its capacity. They may be doing different classes of work, and at any distance from the outfit up to limits fixed by the allowable voltage drop in the lines. This feature is particularly effective in those cases where the job is large enough to permit of several operators working at one time. The low voltage used for welding precludes all chance of accident to the operator from contact with current-carrying parts of the apparatus. Welds made by the electric arc possess a degree of strength only slightly below that of the original section, and by reinforcing this can be increased to any desired amount. They present a neat and finished appearance, are homogeneous in structure and may be easily machined. From every standpoint they are of a highly satisfactory character.

CHAPTER V

ELECTRIC SOLDERING

The application of electricity in welding has been explained in the preceding chapters. As stated, there are two fundamental methods of welding in which the electric current is employed, *i. e.*, the arc and resistance methods. The arc is used to a limited extent for welding large broken parts and its application is considered more economical than any other process, but the danger of handling current at the high voltages that are necessary makes its scope limited. The other welding process, as invented and developed by Dr. Elihu Thomson, consists of causing a heavy current of electricity at a low voltage to flow through the abutting ends of the pieces of metal to be welded. This heats the metal at the joint to a welding temperature.

What is true of welding is also true of the electrical soldering process about to be described, as in both processes heat is developed by the same action, *i. e.*, the passage of a large current of electricity through the joint. This soldering process is a mechanical one and in operation the apparatus used is not likely to give any more trouble than any simple machine will. The wear on the clamping jaws makes it necessary to replace them periodically, but as they are comparatively inexpensive and constitute the only replacement necessary, the operating expense is very low. The amount of the current used in optical framework averages 1 KW. hour per 1500 joints. This current can be purchased from the local lighting company or a generator can be installed which would probably reduce the current expense.

How the Process is Conducted

The general method of soldering consists of holding the pieces to be joined by clamping jaws with the ends of the work in firm contact. A heavy current of electricity, regulated to heat the joint sufficiently to melt the solder, is next passed through the work. The solder, in the form of tape or wire, is then applied to the joint. It flows in and around all parts heated to the proper temperature, as when using a gas flame, but an important difference is noted: the "life" or temper is retained in pieces that have been electrically soldered, instead of their being left in an annealed condition as when heated with a flame. One theoretical reason for this is based on the fact that alternating current electricity travels on the surface of a conductor, and so the core of the work does not heat to a temperature sufficient to become annealed. This condition is illustrated in Fig. 1. The heat varies from a maximum at the joint to the normal temperature of the machine at the jaws, and the heated section would take some such form as shown. As the length of the work that is heated is relatively short, the distance between the clamps usually being twice the diameter of the work, the heat has not had time to run into the work before the joint is made and the current shut off. This is shown by the fact that two highly tempered wires soldered together by the electrical process offer the same resistance to being bent at any other point as at the joint. The yield point or bending strength of the metal is practically as high as before heating.

Range of Electrical Soldering

Practically all of the metals such as brass, copper, steel, German silver, gold, and silver can be soldered successfully in this way, and it is without doubt the most economical method for a continuous run of work. There are no noxious fumes or smoke produced in making an electrically soldered joint, and windows can be opened in warm

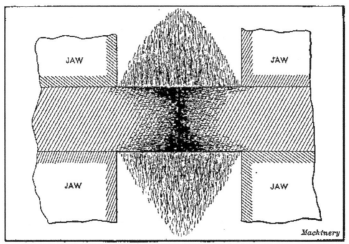

Fig. 1. Diagram showing Relative Volume of the Work that is
heated by the Current

weather without affecting the process in the least. The operator is thus able to do a full day's work every day, instead of experiencing the fatigue that is caused by breathing the carbonic acid gas caused by the gas flames. The joint is made almost instantly, the time required to heat the joint, apply the solder, and shut off the current being approximately from three to five seconds, depending on the cross-sectional area of the joint. As the gripping jaws of the holders are made as large as possible, the heat is drawn from the work almost the instant that the current is shut off, allowing the work to be removed immediately.

Examples of Electrically Soldered Optical Frames

A few samples of parts of eyeglass frames joined by this process are illustrated in Fig. 2. At the extreme right is shown a "cable-temple" before and after the joint is made. These cable ends are wound in a special machine and consist of two coils, right- and left-hand, one inside the other. The inner coil is made of brass wire wound on a

steel wire arbor and then swaged to a specified diameter. The outer
coil is made of German silver, gold filled, or any other stock that is
desired, and it is pushed over the inner coil. After the assembled
cable is soldered to the "temple," which is a solid wire with the center
reduced, it is swaged to the final finished diameter. This leaves a
very smooth and flexible ear-piece, and at the same time a stiff connection to the lens-holders. The soldering of the brass-German-silver
cables caused some trouble through the brass fusing before the German
silver would heat enough to flow the solder, but this was stopped by
using a larger wire in making the secondary coil of the transformer.

Two specimens of "nose-pieces" soldered to "eyes" are shown to the
left of the cable-temples. These eyes are formed by rolling a round

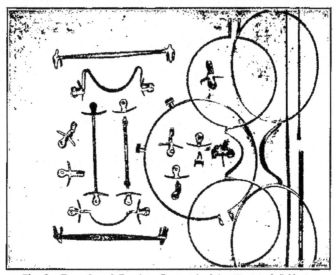

Fig. 2. Examples of Eyeglass Parts joined by Electrical Soldering

wire to form a groove in it; they are then wound on an arbor and
sawed apart. The end-pieces are sawed, assembled and peened in one
machine, and they were formerly soldered by gas. Previous to soldering the "bridges" on by electricity, a long space was annealed on the
eyes. This made a joint that could be easily bent, and various methods
of striking in dies were resorted to in order to get back some of the
temper. In all cases of soldering by electricity, the eye wire is left
with nearly all of the original temper. Another eye with studs attached is shown encircling samples of "straps," "studs" and "end
pieces" before and after assembling and soldering, and to the left of
this eye are shown different forms of bridges and nose-pieces with
straps, before and after soldering. Such parts that have about the
same cross-sectional area at the joint, are very easily handled.

The Utilization of High Voltage Alternating Current

In the process of electrical soldering, alternating current is invari-
ably used, although there is no fundamental reason why direct current

can not be employed. For mechanical and economical reasons, how-ever, direct current is not to be considered. To make this clear, sup-pose a joint having a cross-sectional area of 0.125 square inch re-quires a current of 130 amperes at 3 volts to heat it properly, and that an ordinary plating dynamo rated at these figures is used to furnish the current. It will be noticed that the work heats practically solid from jaw to jaw. Then suppose a joint having a cross-sectional area of one-half the first one, or 0.063 square inch, is to be heated by the same dynamo. A suitable resistance must be interposed in order to reduce the current to a point where the joint will heat properly without melting. This resistance will use current as though it were doing useful work and the small joint will cost practically the same as the large one, as regards the amount of power consumed. On the other hand, it is claimed that the heating action of alternating cur-rent is more uniform, as it flows more on the surface; the heat is thus more intense on the surface and is evenly conducted to the core of the pieces, offsetting the effect of radiation and conductance.

The current used for electrical soldering should be a single phase alternating current of any frequency between 40 and 60. A higher frequency could be used, but it is not good practice for various reasons. A step-down transformer of the shell-core type is preferably used to reduce the 110 or 220 volt feed pressure down to the 1½ to 5 volts required at the machine jaws. It has been found that a pressure of from 1½ to 5 volts is sufficient for all optical frame work, and from 75 to 500 amperes of current is consumed. The use of a large transformer for small work is wasteful, as, although the current can be regulated as desired without much loss of energy, the work heats much more slowly than when a transformer of the proper capacity is used. The machine transformer is usually connected in series with a single phase generator, but it may also be connected to one phase of a polyphase circuit or to either phase of a two phase generator.

The Transformer

The transformer is made by winding a coil of very large insulated copper wire around a core built up of iron sheets cut to shape by dies, each sheet being insulated from the other by shellac or some other medium. This coil, known as the secondary coil, is carefully insulated from the primary coil, which consists of a large number of turns of smaller wire wound around the secondary coil and its core. The num-ber of turns of fine wire depends upon the number of turns of heavy wire and the current to be taken in and given out; also on the rules governing transformer design. The type of transformer illustrated in Fig. 3 is particularly well adapted for use in electrical soldering, as it can be used without changes with other work holders; and this would not be the case if it were built into the machine. As shown, it has the coils protected by an iron cover which not only acts as a case, but also as part of the magnetic field. Transformers of this type are very efficient—from 95 to 97 per cent of the current taken in being given out—and they are particularly suited for constant work.

Unit System of Electrical Soldering

The writer has developed a "unit system" of soldering and applied it very successfully in the manufacture of optical frames. This system consists in mounting all the working parts of the machine for each particular operation on a base-board or stand. Figs. 3 and 4 illustrate this idea; the transformer is mounted at the center, with a fuse box at the rear and the work holder at the front of the board. Under the base-board is located the adjustable rheostat operated by a sliding plate shown at the side. To set up this machine at any position in the shop, it is only necessary to run two wires from the feed circuit and attach a foot treadle to operate the clamp jaws and switch. This system allows the same transformer and other parts to be used with another machine in case of a change or the discarding of the original machine.

There are two practical methods of controlling the heat obtained at the joint; one is by introducing an adjustable rheostat into the

Fig. 3. Example of Unit Equipment for soldering Optical Frames

primary circuit, as illustrated; and the other method is to introduce a reactive or "choke" coil into the same circuit. Of the two, the reactive coil is undoubtedly the better, as there is practically no loss of power and an infinite number of adjustments may be made, whereas the rheostat is limited to the number of contact points used. The difference in loss of current is an inappreciable amount more with the rheostat, but it can be made for little expense and for that reason has been used more than the choke coil. The writer uses the rheostat control on nearly all of his equipments on account of its simplicity, the ease with which it may be built and the simplicity of operation.

A machine for soldering straps to eye-pieces and bridges is shown in Fig. 5. This machine or holder consists of a base A with a vertical slide B working in a slot at the rear. A second slide C also works in another slot at the rear, the slot being inclined at 45 degrees to the base. This slide C is operated through a lever D which receives its movement from the slide B; the lever D is pivoted in the base. The slide B is provided with a spring tension which allows the lever D to keep a constant pressure on the slide C while the slide B continues to move. The lever D works in a slot cut through the slide C, this slide carrying a cam-operated swinging arm at its upper end to which the clamping jaws G are attached. This upper jaw is designed to swing away from the work and leave it clear to facilitate handling.

At the rear of the machine and attached to the base there is a switch which is operated by a pin in the slide B. At the front of the base

Fig. 4. Closer View of Work-holding Jaws shown in Fig. 3

and insulated from it is the casting I which is milled to receive an arm L that is free to move on a pivot, but the motion of the arm is limited by the adjusting screws J and K. The arm is held against the screw J by means of an adjustable spring tension N. There is a jaw O at the upper end of the arm, which, in this case, holds the strap in the proper relation to the other part to which it is to be soldered. The contact of the jaw O with this strap is made by the pressure of the spring N against the arm L, and the strap is held against the part to which it is to be soldered, which is carried between the jaws P and G. The jaws are made interchangeable for different classes of work.

At the lower end of the casting I, one end of the secondary or low pressure circuit is connected by means of the terminal T, and a spring brush R is used to insure a low resistance contact between the casting I and the rocker arm L. The lower clamping jaw P is attached to the base A and the jaw is provided with a gage for aligning the part held in it. The slide B is held at the top of its movement by means of the spring S. Two points of the switch control the primary or high pressure circuit, and the other two points operate on the secondary which is in the circuit with the jaws of the machine. A chain connects the lower part of the slide B with a foot treadle which is placed under the bench in a convenient position for the operator.

The operation of this holder is as follows: Two pieces to be joined, previously covered with a non-scaling or protective mixture, have the

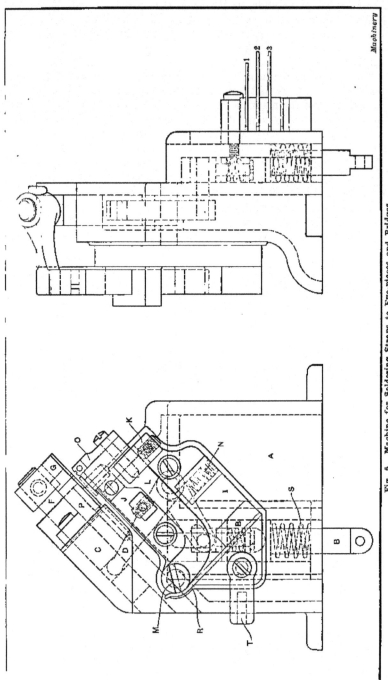

Fig. 5.   Machine for Soldering Straps to Eye-pieces and Bridges

joint end of one piece dipped into the flux. They are then assembled in the proper relation to each other in the jaws of the holder, which are so arranged that the rocking arm is away from its stop when the work is in place. The foot treadle next is depressed until the upper clamp jaw grips the work; in this case only one part is held rigid. The other piece—which is a strap—is guided by its form and a teat on the piece held in the rigid jaws. The solder, in the form of wire, is then placed on the junction and the foot lever depressed further until the current is connected. Almost instantly the solder ·flows and runs around the joint, when the foot treadle is released entirely, and the work, which is left free, is taken out with a pair of tweezers. On work which is very small and difficult to handle with the fingers, tweezers are used; but such work as soldering temples together, bridges to eyes, bridges to straps, or eyes to studs, is handled with the fingers. The heat is held at the joint instead of spreading as it does when heated with the flame, so it causes the

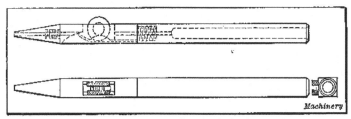

Fig. 6. Holder for Applying the Solder by Hand

operator no discomfort to take the joined pieces out as soon as the jaws are opened. The jaws are brushed clean at intervals, using a short hair stiff bristle brush for this purpose.

The idea of using a spring tension jaw was developed by the writer after having had considerable trouble caused by particles of dirt or burrs getting into the junction, also by not having the two ends fit together properly to form a contact of low resistance. By the movable jaw, all of this trouble was eliminated as the constant spring pressure holds the ends in firm contact, automatically keeps the ends together in the case of burrs or other points fusing, and prevents any break in the contact while the current is being applied. In the welding process, the ends are forced together while at a welding temperature, but this changes the form of the ends and shortens the pieces; consequently it could not be applied to optical work, as there must be no change in the size or form of the pieces to be joined. The spring behind the rocking arm *L* in Fig. 5 is adjusted to provide just sufficient tension to keep a constant pressure on the junction without deforming or upsetting the ends, thus forming the joint when the ends become hot. The jaws of the holder are made as large and heavy as possible to allow of their working continuously without heating. These jaws are made of copper, which has been found best for this purpose on account of the low resistance of the contact made between them and the metal to be operated on.

Preparing the Work to Prevent Scaling

To prevent gold-filled metal from scaling or "burning" at the joint, it is customary to cover the work with some preparation to prevent oxidation. Probably the best, and at the same time the simplest, method of preparing the work is to place it in an ordinary flour sieve, cover it with commercial boracic acid, and then shake all loose powder out. This leaves the parts covered with a thin coating of dust which becomes liquid at a low red heat and prevents the air from coming into contact with the surface of the gold. Another method is to make a solution of the boracic acid and water, dip the pieces into this solution, drain off the surplus and allow them to dry. In no case, however, should any solution be used that will leave a hard film over the parts, as this would prevent a clean contact with the clamping jaw, create a resistance that would cause an arc to develop, and spoil the surface of the work. The flux generally used is borax and it is prepared in the following manner: A piece of genuine slate, the green colored variety, which is the hardest, being the best, is thoroughly cleaned; a few drops of water are placed in the center and a thick, creamy mixture of borax is made by rubbing a piece of crystalline borax in the water on the slate until the desired consistency is obtained. The proper mixture is best determined by actual trial; a mixture that is too thin or too thick will either cause the solder to remain in one spot, instead of flowing through the joint, or create an unclean contact and interfere with the heating.

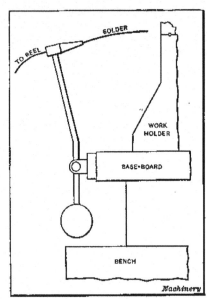

Fig. 7.   Bracket for supporting Solder Holder on the Machine

The solder in the form of wire may be held in the hand in a holder, as shown in Fig. 6, or some such arrangement as the one shown in Fig. 7 may be employed. This consists of a chuck at the top of a wire, bent about as shown, and having a metal ball at the lower end heavy enough to balance the wire and chuck in an upright position. This wire is held by a screw in one member of a universal joint which allows the chuck to be moved freely to any position in front of the clamping jaws and take a convenient position to allow the solder to be grasped by the operator. When a holder of this type is used, both hands are free to place and adjust the work and apply the solder quickly.

Lightning Source UK Ltd.
Milton Keynes UK
UKOW04f0750301117
313622UK00002B/244/P